My Career in Clinical Research

My Career in Clinical Research

John Battish, MS

Copyright © 2024 by John Battish, MS.

ISBN: Softcover 979-8-3694-2932-7
       eBook   979-8-3694-2931-0

All rights reserved. No part of this book may be reproduced or transmitted in any form or by any means, electronic or mechanical, including photocopying, recording, or by any information storage and retrieval system, without permission in writing from the copyright owner.

Any people depicted in stock imagery provided by Getty Images are models, and such images are being used for illustrative purposes only.
Certain stock imagery © Getty Images.

Print information available on the last page.

Rev. date: 09/24/2024

To order additional copies of this book, contact:
Xlibris
844-714-8691
www.Xlibris.com
Orders@Xlibris.com
819513

Contents

Chapter 1   Medicine comes from the heart............. 1
Chapter 2   My Years at Stellar – Chance at University of Pennsylvania (1999-2001) 4
Chapter 3   Jesse Gelsinger Impact on Gene Therapy... 9
Chapter 4   My Years Working in the Emergency Room............................... 14
Chapter 5   Penn Abramson Cancer Center 19
Chapter 6   Sanofi Aventis 24
Chapter 7   OXALIPLATIN (Brand name – Eloxatin)... 28
Chapter 8   United States Food and Drug Administration / International Clinical Trials 32
Chapter 9   The Clinical Research Associate - Monitoring roles 39
Chapter 10 Clinical Research Organization............. 43
Chapter 11 My years working for KFORCE (2007-2011).. 47
Chapter 12 My years at MORPHOTEK (2012-2015).... 54

Chapter 13	Merck and Keytruda Years (2015-2017)	58
Chapter 14	Chiltern, Covance, and Labcorp Years (2017-2022)	68
Chapter 15	Approval of Yescarta	76
Chapter 16	Reasons and Importance of Clinical Research Monitoring	80

References 85

*"Dedicated to my parents
Jangeshwar Battish
and Veena Battish"*

CHAPTER 1

Medicine comes from the heart

I STOOD IN front of my patient at the Hospital of the University of Pennsylvania. It was the year 1999. I had a list of clinical questions to ask my patient as part of my hospital duties. However, I then noticed the personal diary he was keeping close to himself. My patient then told me he was determined to conquer his cancer. This scene was the beginning to my experiences in clinical research. Through later years, I eventually helped to contribute to the FDA approvals of cancer treatment drugs such as Oxalitin, Sutent, Keytruda, and Yescarta. After I completed my clinical research years at the Hospital of University of Pennsylvania in 2002, I spent seventeen years working for various pharmaceutical companies. I worked with Sanofi-Synthelabo, Pfizer, Abbott, Morphotek, Merck, and Kite. I have seen great advancements in clinical research while working with these companies. The aim of this book is to provide guidance for individuals who wish to work in the exciting field of clinical research. This book will also provide the proper steps to become an expert in clinical trials research. The reader will gain a general knowledge of clinical research trials and how they are conducted. The reader will also learn the background of the various pharmaceutical companies and agencies working on these clinical research trials.

I first graduated in 1997 from the University of Delaware biology of science degree. I did not really know what I wanted to do in life afterwards. I had originally planned to become a medical doctor. However, I eventually decided I wanted to go into medical research along with general lab research. I took a volunteer position at the Alfred AI DuPont Children's Hospital in Wilmington DE. I worked with a Doctor Lendel and his research staff assistance in the summer of 1997. It was in his lab that I learned about DNA synthesis and the use of gels to do DNA research. I also learned about using tubes and centrifuges for breaking down DNA for further studies. How then did I eventually get into working for medical clinical research trials? In this book, you will begin to understand how the progression of DNA recombinant technology led me to the exciting field of clinical research for combating medical diseases through targeted DNA medication. In 2008, I even received a Masters of Science in Clinical Research Organization and Management from the Drexel College Medical University. This book is a description of my career journey in medical clinical research from 1999-2022.

CHAPTER 2

My Years at Stellar – Chance at University of Pennsylvania (1999-2001)

THE UNIVERSITY OF Pennsylvania (UPenn) is a prestigious Ivy League institution located in Philadelphia, Pennsylvania, USA. It is renowned for its contributions to various fields of research and academia, including medical and scientific research.

UPenn's Perelman School of Medicine is known for its cutting-edge research in various medical disciplines. Within the Perelman School of Medicine, there are several research laboratories, departments, and centers dedicated to advancing knowledge in different areas of medicine and biomedical sciences.

I first started at Chellar Stance Laboratories at the University of Pennsylvania. I was given the position as the main Gene Therapy Lab Manager. It was here that I learned further about the developing concepts of gene therapy. This was the year 1999 so gene therapy was just in its infancy. Gene therapy would at first have its troubles in its growth. Eventually later in the 2000s, gene therapy would become a major force of clinical research. I did not know at the time how gene therapy would become very important in the field of medicine. Especially since gene therapy started out with major issues in the early 21st century.

Gene therapy is a cutting-edge medical approach that aims to treat or cure genetic diseases by modifying or replacing faulty genes with healthy ones. It is based on the understanding that many diseases are caused by specific genetic mutations or abnormalities, and by targeting and correcting these genetic defects, it is possible to treat the underlying cause of the condition.

The basic concept of gene therapy involves the introduction of genetic material into a patient's cells to either replace a faulty gene or provide a functional copy of the gene. This genetic material can be delivered using different methods, such as viral vectors, non-viral vectors, or other techniques.

There are two main types of gene therapy:

1. Somatic Gene Therapy:
   Somatic gene therapy targets the non-reproductive (somatic) cells of the body. The genetic modifications made through somatic gene therapy are not passed on to future generations. This approach is used to treat genetic disorders and certain acquired diseases, such as certain types of cancer. The corrected genes are usually delivered directly to the affected tissues or organs, where they can produce the functional proteins needed to treat the disease.

2. Germline Gene Therapy:
   Germline gene therapy aims to modify the genes in reproductive cells (germ cells) or early-stage embryos. Unlike somatic gene therapy, the genetic changes made in germline

gene therapy can be passed on to future generations. This approach has the potential to prevent genetic diseases from being inherited by descendants, but it raises significant ethical and safety concerns. As of my last update in September 2021, germline gene therapy was not considered safe or ethical for use in humans due to the risk of unintended consequences and ethical considerations.

Gene therapy has shown promising results in the treatment of certain genetic disorders, such as severe combined immunodeficiency (SCID), also known as "bubble boy" disease, and certain types of inherited blindness. Additionally, ongoing research and clinical trials are exploring gene therapy's potential for treating various genetic disorders, cancer, and other complex diseases.

Challenges and Considerations for gene therapy:

Gene therapy is a complex and evolving field, and it comes with some challenges and considerations, including:

1. Delivery: Ensuring that the corrected genes reach the appropriate cells and tissues in the body can be challenging.
2. Safety: Gene therapy must be carefully designed and tested to avoid unintended consequences, such as off-target effects or an immune response against the viral vectors used for delivery.
3. Ethical Issues: Germline gene therapy raises significant ethical concerns, as it involves making heritable changes to future generations.
4. Long-term effects: Monitoring the long-term safety and efficacy of gene therapy treatments is essential, as the effects may evolve over time.

Despite these challenges, gene therapy holds great promise for revolutionizing the treatment of genetic diseases and offering potential cures for conditions that were previously considered untreatable. As the field advances, ongoing research and clinical trials will continue to shape the future of gene therapy and its potential applications in medicine.

CHAPTER 3

Jesse Gelsinger Impact on Gene Therapy

WHILE I WAS working at Stellar Chance laboratory, a young man named Jesse Gelsinger passed away at the Hospital of the University of Pennsylvania. Jesse Gelsinger was a young man who tragically died in 1999 during a clinical trial for gene therapy, marking a significant milestone in the history of gene therapy. His death had a profound impact on the field, leading to increased scrutiny, reassessment of safety protocols, and heightened ethical considerations in gene therapy research.

Jesse Gelsinger's Case:

Jesse Gelsinger was an 18-year-old with a rare genetic disorder called ornithine transcarbamylase (OTC) deficiency. OTC deficiency affects the body's ability to process ammonia, leading to toxic levels of ammonia buildup in the blood. In September 1999, he volunteered to participate in a gene therapy clinical trial at the University of Pennsylvania. The trial aimed to use a genetically modified virus to deliver a functional copy of the OTC gene into his liver cells, potentially correcting the underlying genetic defect.

Tragically, shortly after receiving the gene therapy treatment, Jesse experienced a severe immune response known as a cytokine storm. This reaction led to multiple organ failure, and despite intensive medical efforts, Jesse passed away just a few days after the gene therapy administration.

Impact on Gene Therapy:

Jesse Gelsinger's death had a profound impact on the field of gene therapy, triggering several significant changes:

1. Increased Safety Oversight: The tragic outcome of Jesse's case highlighted the need for more rigorous safety measures in gene therapy trials. Researchers, regulators, and institutions reevaluated the protocols for administering gene therapies, with a particular focus on avoiding severe immune reactions and other adverse events.
2. Regulatory Changes: The incident prompted increased scrutiny from regulatory agencies, such as the U.S. Food and Drug Administration (FDA). The FDA implemented stricter guidelines for gene therapy trials, requiring more comprehensive preclinical testing, careful patient selection, and more detailed reporting of adverse events.
3. Ethical Considerations: Jesse's death sparked discussions about the ethical implications of conducting gene therapy trials, especially in cases where the volunteers have severe medical conditions and may not fully comprehend the potential risks involved.
4. Public Perception: The incident garnered significant media attention and public concern

about the safety of gene therapy. This led to a temporary slowdown in gene therapy research as public trust needed to be rebuilt.
5. Improved Science and Research: Despite the setback, Jesse Gelsinger's case led to an increased focus on understanding the immune responses triggered by gene therapies. Researchers worked to develop safer vectors for gene delivery and explore methods to prevent adverse reactions.

Over the years, gene therapy research has continued to advance, benefiting from the lessons learned from Jesse Gelsinger's case. Today, gene therapy is a rapidly developing field, and several gene therapies have been successfully approved for various genetic diseases, such as spinal muscular atrophy and certain types of inherited blindness. Researchers and regulators remain vigilant in prioritizing patient safety and ethical considerations while pushing the boundaries of this promising medical approach.

However due to the Jessie Gelsinger case issues just after they broke out, I was forced to find another field of study as gene therapy was put on hold while its issues were being resolved. In the year 2000, I started working in clinical research in the hospital setting at the University of Pennsylvania. I decided to start clinical research in the emergency department at the Hospital of the University of Pennsylvania.

CHAPTER 4

My Years Working in the Emergency Room

I WILL FIRST provide a description of the emergency room that I worked for at the University of Pennsylvania. This will be the area which I collected a lot of data for clinical research purposes. An emergency room (ER) in a hospital is a specialized medical facility designed to provide immediate medical care to patients with acute and potentially life-threatening conditions. Also known as the emergency department (ED), the ER is staffed by a team of healthcare professionals trained to handle various medical emergencies. Here's a description of what you might find in an emergency room:

1. Triage Area: The ER typically has a triage area where specially trained nurses or medical personnel assess patients as they arrive. Triage is a process of prioritizing patients based on the severity of their condition. Patients with life-threatening conditions or severe injuries are given priority and receive immediate attention.
2. Treatment Rooms: The ER consists of various treatment rooms or bays equipped to provide medical care to patients. These rooms are equipped with essential medical equipment and supplies to handle a wide range of emergencies, from minor injuries to critical medical conditions.
3. Medical Staff: The ER is staffed 24/7 by a team of medical professionals, including emergency medicine physicians, nurses, paramedics, and other healthcare staff. These professionals have specialized training to handle medical emergencies and respond quickly to critical situations.

4. Medical Equipment: The ER is equipped with a range of medical equipment to diagnose and treat patients efficiently. This equipment may include cardiac monitors, defibrillators, infusion pumps, ventilators, imaging equipment (X-ray, CT scan), and laboratory facilities for rapid blood tests.
5. Resuscitation Area: Many ERs have a designated resuscitation area, also known as the trauma bay, where critical patients with life-threatening conditions are treated. This area is equipped with advanced life support equipment and is staffed by a specialized team trained to manage emergencies.
6. Waiting Area: Most ERs have a waiting area where patients or their family members wait for assessment and treatment. The waiting area may have triage nurses available to assess patients' conditions and determine the order of treatment.
7. Security: ERs often have security measures in place to ensure the safety of patients, staff, and visitors, especially during chaotic situations or high-stress moments.

8. Support Services: The ER may have access to support services such as social workers, interpreters, chaplains, and mental health professionals to assist patients and their families during difficult situations.
9. Access to Specialty Care: In larger hospitals, the ER may have access to various specialty care teams, such as neurologists, cardiologists, surgeons, and pediatricians, who can be consulted for specialized care if needed.

Emergency rooms are designed to provide immediate medical care and stabilization to patients in crisis. They serve as a crucial entry point to the healthcare system for individuals experiencing medical emergencies and are an essential part of hospitals' overall healthcare services.

My job in the emergency rooms was to gather data that patients gave permission to be used for clinical research studies. The physicians would use this gathered clinical research data to see if there were any noticeable trends. We were able to notice various trends in follow-up for treatment of subjects in the city. Most subjects were eventually lost to follow-up due to broken communication or inaccurate contact information being collected. This allowed the Emergency Room workers to ensure patients were providing accurate follow-up information. I also worked for the Children's Hospital of the University of Pennsylvania. I ensured the contact information being collected at the Children's Hospital was accurate. I worked on these emergency department studies for a few months. However, I was then offered a position at the Abramson Cancer Center at the University of Pennsylvania. My journey into cancer treatment clinical research trials would soon begin.

CHAPTER 5

Penn Abramson Cancer Center

LET US FIRST look at the general background of the cancer center that I worked at. It was a cancer center where in the mornings in its lobby I would remember getting Cheese Danishes and coffee before heading up to the office area for work. The Abramson Cancer Center at the University of Pennsylvania (ACC) has a rich and distinguished history in cancer research, treatment, and patient care. It is one of the leading cancer centers in the United States and has made significant contributions to the field of oncology. Here is an overview of its history:

1. Founding: The Abramson Cancer Center was founded in 1973 as the "Cancer Research Institute" at the University of Pennsylvania. It was established to focus on cancer research, education, and patient care. In 1991, Leonard and Madlyn Abramson made a transformative gift to the center, and it was renamed the "Abramson Cancer Center" in their honor.
2. Accreditation: The ACC received its first designation as a Comprehensive Cancer Center by the National Cancer Institute (NCI) in 1973. This prestigious designation signifies that the center meets stringent criteria in cancer research, clinical care, and community outreach.
3. 3. Research and Innovation: Over the years, the ACC has been at the forefront of groundbreaking cancer research and innovations. Its researchers have made significant contributions in areas such as immunotherapy, targeted therapies, cancer genetics, and cancer prevention.

4. Patient Care: The ACC provides comprehensive, multidisciplinary cancer care to patients with a wide range of cancer types. It offers state-of-the-art treatments, personalized therapies, and supportive care services to improve patients' quality of life.
5. Collaborations: The ACC collaborates with other institutions, both nationally and internationally, to advance cancer research and share knowledge. These collaborations have facilitated significant advancements in cancer care and treatment options.
6. Translational Research: The ACC places a strong emphasis on translational research, bridging the gap between laboratory discoveries and clinical applications. This approach has allowed researchers to rapidly translate new scientific findings into potential treatments for patients.
7. Community Outreach: The ACC is actively involved in community outreach and education programs. It works to raise awareness about cancer prevention, early detection, and the importance of clinical trials.

8. Continued Growth: Over the years, the ACC has expanded its facilities, programs, and expertise. It has attracted world-class researchers, physicians, and healthcare professionals dedicated to advancing cancer care and research.

Today, the Abramson Cancer Center at the University of Pennsylvania continues to be a leader in cancer research and patient care. It is committed to advancing knowledge, developing new treatments, and providing compassionate care to cancer patients and their families. Its contributions to the field of oncology have had a lasting impact on cancer care both within the University of Pennsylvania community and beyond.

I worked at the Abramson Cancer Center for two years. It was here that I was able to see the determination of the patients in fighting for their survival. It energized and encouraged me to use my gained knowledge to become a medical monitor. I had a mentor at the Abramson Cancer Center who taught me all about the medical monitoring position. One of the cancer drugs being developed in 2002 was the cancer treatment drug known as Eloxatin. It was a drug being developed by Sanofi-Aventis. In the year 2004, I started a job position at Sanofi-Aventis in Malvern, PA in their international clinical trials department. I worked in their international medical records division.

CHAPTER 6

Sanofi Aventis

SANOFI-AVENTIS, NOW KNOWN simply as Sanofi, is a multinational pharmaceutical company which is mainly headquartered in Paris, France. The company has a long and rich history, with its roots dating back to the 19th century. This company would become a major contributor to cancer medical clinical trials and research. Here's an overview of the history of Sanofi:

Founding of Sanofi:

Sanofi was founded in 1973 through the merger of two French pharmaceutical companies, Sanofi and Synthélabo. The newly formed entity was named Sanofi-Synthélabo.

Sanofi:

The original Sanofi company was founded in 1973, but its history traces back to 1970 when Elf Aquitaine, a French energy company, established a pharmaceutical division. This division eventually became Sanofi. In its early years, Sanofi focused on developing and marketing a wide range of pharmaceutical products.

Synthélabo:

Synthélabo was established in 1970 as a spin-off from the French pharmaceutical company Dausse. It also started as a pharmaceutical division and grew rapidly through the development and commercialization of innovative medicines.

Merger and Formation of Sanofi-Synthélabo:

In 1999, Sanofi and Synthélabo merged to form Sanofi-Synthélabo, creating one of the largest pharmaceutical companies in France and Europe. The merger brought together a diverse portfolio of medicines and research expertise from both companies.

Aventis:

In 2004, Sanofi-Synthélabo acquired Aventis, a German-French pharmaceutical company that was formed through the merger of Hoechst Marion Roussel and Rhône-Poulenc Rorer in 1999. The acquisition was finalized in 2004, and the merged entity was named Sanofi-Aventis.

Sanofi:

In 2011, the company simplified its name to Sanofi, dropping the "Aventis" part from its corporate identity. The name change was aimed at streamlining the company's brand and corporate image.

Diversification and Expansion:

Throughout its history, Sanofi has focused on diversifying its business and expanding its global presence through acquisitions, collaborations, and partnerships. The company has expanded its portfolio to include a wide range of therapeutic areas, such as cardiovascular, central nervous system, vaccines, oncology, rare diseases, and more.

Continued Innovation and Research:

Sanofi has maintained a strong commitment to research and development, investing heavily in the discovery and development of new medicines and therapies. The company has established research centers worldwide and collaborates with academic institutions and biotechnology companies to advance medical science.

Sanofi continues to be a major player in the pharmaceutical industry, with a global presence and a strong portfolio of medicines.

CHAPTER 7

OXALIPLATIN (Brand name – Eloxatin)

THE MAJOR CANCER drug developed during my years working at the Abramson Cancer Center and eventually my job at Sanofi-Aventis was the cancer treatment drug now known as Eloxatin. This drug was first known under its study drug name of Oxaliplatin. The study drug Oxaliplatin is an anticancer medication manufactured from Sanofi-Synthelabo that belongs to the class of platinum-based chemotherapy drugs. It is used to treat various types of cancer, most notably colorectal cancer. As stated before, Oxaliplatin is now sold under the brand name Eloxatin.

Mechanism of Action:

Oxaliplatin exerts its anticancer effects through a unique mechanism of action. It is a platinum coordination complex that forms DNA adducts, which are abnormal links between the drug and DNA strands. These adducts interfere with DNA replication and transcription, ultimately leading to cell death. Oxaliplatin's ability to damage DNA makes it effective against rapidly dividing cancer cells, which are particularly susceptible to DNA disruption.

Indications:

Oxaliplatin is primarily used in combination with other chemotherapy agents, such as 5-fluorouracil (5-FU) and leucovorin, to treat advanced or metastatic colorectal cancer. It is also used in adjuvant therapy (after surgical removal of the tumor) for certain stages of colorectal cancer to reduce the risk of cancer recurrence.

Oxaliplatin has shown some efficacy in treating other cancers, including certain types of gastric cancer and pancreatic cancer. Its use is continuously studied in clinical trials for various malignancies.

Administration and Dosage:

Oxaliplatin is administered intravenously (IV) in a clinical setting, typically as an infusion into a vein in the arm. The dosage and treatment schedule may vary based on the specific cancer being treated, the patient's overall health, and other treatment modalities used in combination with Oxaliplatin.

Side Effects:

Like many chemotherapy drugs, Oxaliplatin can cause side effects. Some common side effects include:

1. Nausea and vomiting
2. Peripheral neuropathy (tingling or numbness in the hands and feet)
3. Fatigue
4. Diarrhea or constipation
5. Cold sensitivity, especially in the hands and throat

6. Myelosuppression (reduction in blood cell counts), which may lead to an increased risk of infections, anemia, and bleeding.

It is essential for patients to be closely monitored during treatment, and their healthcare team may recommend measures to manage side effects and improve overall comfort and quality of life.

Precautions:

Oxaliplatin should be used under the supervision of a qualified healthcare provider experienced in administering chemotherapy. Patients with a history of severe allergic reactions to platinum-based drugs or similar medications should not receive Oxaliplatin.

As with any chemotherapy, potential benefits should be weighed against potential risks and side effects. Patients should discuss their medical history and current health status with their oncologist before starting Oxaliplatin treatment.

CHAPTER 8

United States Food and Drug Administration / International Clinical Trials

I WAS ENCOURAGED by my mentors in 2005 to become a medical monitor for major pharmaceutical companies. Medical monitoring is used to ensure the clinical research data collected was accurate and kept within the standards of the Food and Drug Administration.

Role of the FDA

The Food and Drug Administration (FDA) is a regulatory agency of the United States Department of Health and Human Services. Its primary role is to protect and promote public health by regulating and supervising various products, including food, drugs, medical devices, cosmetics, tobacco, and more. The FDA's responsibilities extend to ensuring the safety, efficacy, and security of these products, as well as advancing public health initiatives and disseminating accurate health information. Here are the main roles and functions of the FDA:

1. Drug Regulation: One of the FDA's core responsibilities is to regulate pharmaceutical drugs. Before a new drug can be marketed and sold in the United States, it must undergo rigorous testing through preclinical studies and clinical trials. The FDA reviews the data from these trials to determine whether the drug is safe and effective for its intended use. If the drug meets the agency's standards, it will be approved for commercial distribution.

2. Medical Device Regulation: The FDA oversees the safety and effectiveness of medical devices, ranging from simple tools like tongue depressors to complex devices such as pacemakers and artificial joints. Similar to drugs, medical devices must undergo evaluation and clearance or approval before they can be legally marketed in the United States.
3. Food Safety: The FDA is responsible for ensuring that the nation's food supply is safe and properly labeled. It establishes regulations and inspects food facilities to prevent contamination, outbreaks, and improper handling of food products.
4. Cosmetics and Personal Care Products: The FDA regulates cosmetics and personal care products to ensure they are safe for consumers. This includes overseeing labeling requirements and ingredients used in these products.
5. Biologics: The FDA also regulates biologics, which are medical products derived from living organisms, such as vaccines, blood products, and gene therapies. Similar to drugs, biologics

must undergo thorough testing and approval processes before they can be used for medical treatment.
6. Tobacco Products: In 2009, the Family Smoking Prevention and Tobacco Control Act granted the FDA authority to regulate tobacco products, including cigarettes, cigars, and smokeless tobacco, with the aim of reducing tobacco-related harm.
7. Advancing Public Health: The FDA engages in various public health initiatives, such as promoting healthy behaviors, providing accurate health information to the public, and supporting research to address critical health issues.
8. Emergency Response: In times of public health emergencies, such as outbreaks, natural disasters, or bioterrorism threats, the FDA plays a crucial role in coordinating responses, providing guidance, and expediting the availability of necessary medical products.

The FDA's work involves a delicate balance between ensuring the availability of safe and effective products for consumers and promoting innovation in the fields of medicine and healthcare. Through its regulatory oversight, the FDA seeks to protect public health and uphold the highest standards of safety and quality for the products it oversees.

I also want to briefly mention international clinical trials specifically the European Union in this section:

The European Union (EU) has a structured regulatory framework for international clinical trials designed to ensure safety, efficacy, and ethical standards.

Here's a brief summary for EU Clinical Trials Regulation:

1. **Clinical Trials Regulation (CTR):** The primary legislation is the EU Clinical Trials Regulation (EU) No 536/2014, which aims to streamline the approval process and increase transparency. This regulation governs the conduct of clinical trials across EU member states and the European Economic Area (EEA).
2. **Centralized Application:** Sponsors submit a single application through the Clinical Trials Information System (CTIS), which facilitates

a coordinated assessment by all involved member states. This centralized process aims to reduce administrative burden and accelerate trial approval.

3. **Ethical Review:** Each trial must undergo ethical review by a competent ethics committee or Institutional Review Board (IRB) in the member state where the trial is conducted. This ensures that the trial meets ethical standards and protects participants' rights.
4. **Regulatory Approval:** In addition to ethical approval, regulatory authorities in each member state where the trial will take place must grant authorization. They assess the trial's design, methodology, and safety measures.
5. **Informed Consent:** Trials must adhere to strict informed consent requirements. Participants must be fully informed about the trial's nature, risks, and benefits before agreeing to participate.
6. **Safety Reporting:** Sponsors are required to report serious adverse events and unexpected side effects to regulatory authorities and ethics committees promptly. Ongoing safety

monitoring is crucial to ensure participant well-being.
7. **Data Protection:** Trials must comply with the EU General Data Protection Regulation (GDPR), ensuring that participants' personal data is handled with confidentiality and respect.
8. **Transparency and Reporting:** Results of clinical trials must be published in a publicly accessible database to promote transparency and contribute to the scientific community.

These regulations aim to harmonize trial practices across Europe, facilitate the conduct of high-quality research, and ensure participant safety and data integrity.

CHAPTER 9

The Clinical Research Associate
- Monitoring roles

I BEGAN MEDICAL monitoring as a CRA. A Clinical Research Associate (CRA) plays a crucial role in the field of clinical research, serving as a key link between the pharmaceutical or biotechnology company sponsoring a clinical trial and the study sites where the trials are conducted. CRAs are responsible for ensuring that clinical trials are conducted in accordance with applicable regulations, guidelines, and protocols, while safeguarding the rights and well-being of study participants. This clinical research role has become very important over the years in helping prove certain study drugs are indeed effective in the treatment of various diseases.

The job role of a Clinical Research Associate involves various responsibilities, including:

1. Site Selection and Initiation: CRAs are involved in the selection and evaluation of potential study sites. They conduct site feasibility assessments to ensure that the sites have the necessary resources, qualified personnel, and patient population to conduct the clinical trial. Once a site is selected, CRAs participate in the initiation of the study at the site, which includes training site staff on the trial protocol, study procedures, and regulatory requirements.
2. Monitoring: One of the primary responsibilities of a CRA is to conduct on-site monitoring visits to the participating study sites. During these visits, CRAs review and verify data collected from study participants, ensuring accuracy and completeness. They also assess whether the study is being conducted in compliance with

the protocol, regulations, and Good Clinical Practice (GCP) guidelines.
3. Protocol Compliance: CRAs closely monitor the study protocol's adherence to ensure that study sites are following the study plan and procedures as outlined in the protocol. They work with site personnel to address any deviations or discrepancies found during monitoring visits.
4. Safety and Ethics Oversight: CRAs are responsible for monitoring participant safety during the trial. They review adverse events and serious adverse events reported by the site and ensure that appropriate measures are taken to protect participants' well-being. CRAs also ensure that the study is conducted in an ethical manner and that participants' informed consent is appropriately obtained.
5. Data Management: CRAs are involved in data collection, validation, and entry processes. They ensure that all essential documents, such as case report forms (CRFs), source documents, and regulatory records, are maintained in compliance with the study requirements.

6. Communication and Collaboration: CRAs serve as the main point of contact between the study site and the sponsor company. They maintain regular communication with site staff and study coordinators, providing guidance and support as needed. Additionally, CRAs collaborate with other members of the study team, such as project managers and clinical trial associates.
7. Regulatory Compliance: CRAs ensure that the clinical trial adheres to regulatory requirements set forth by health authorities, such as the U.S. Food and Drug Administration (FDA) or the European Medicines Agency (EMA). They assist in preparing for regulatory inspections and audits.

Overall, Clinical Research Associates play a critical role in ensuring the integrity, quality, and compliance of clinical trials, ultimately contributing to the advancement of medical knowledge and the development of new therapies and treatments. The job requires strong attention to detail, communication skills, and the ability to work both independently and collaboratively in a dynamic and fast-paced environment.

CHAPTER 10

Clinical Research Organization

MY FIRST JOB as a CRA in 2005 was for a clinical research organization named PAREXEL International. PAREXEL International Corporation is a contract research organization (CRO) that provides services to the biopharmaceutical and medical device industries. It was founded in 1982 by Josef H. von Rickenbach and Anne B. Sayigh in Boston, Massachusetts, USA.

Here's an overview of the history of PAREXEL:

Founding and Early Years:

PAREXEL was established with the vision of offering specialized expertise and services to support drug development and clinical trials for pharmaceutical and biotechnology companies. Josef H. von Rickenbach, the company's co-founder and CEO, played a key role in shaping the company's direction and growth.

Growth and Expansion:

Throughout the 1980s and 1990s, PAREXEL experienced significant growth and expanded its services globally. The company established offices in various countries and developed a strong presence in key pharmaceutical markets. As the demand for clinical research and drug development services increased, PAREXEL became a leading player in the CRO industry.

Diverse Services and Offerings:

PAREXEL expanded its service offerings beyond traditional clinical trial management to include a wide range of services across the drug development lifecycle. These services include clinical trial design, regulatory consulting, data management, biostatistics, pharmacovigilance, market access, and health economics research.

Acquisitions and Collaborations:

PAREXEL's growth was accelerated through strategic acquisitions and collaborations. The company acquired several other CROs and specialized service providers, which helped broaden its capabilities and geographical reach. Notable acquisitions include ClinPhone, a provider of electronic data capture (EDC) and randomization services, in 2008.

IPO and Public Listing:

In 1995, PAREXEL went public with its initial public offering (IPO) on the NASDAQ stock exchange under the ticker symbol "PRXL." The IPO provided the company with additional capital to further expand its operations and services.

Continued Innovations and Technology:

PAREXEL has consistently invested in innovative technologies and processes to enhance its service offerings and improve efficiency in clinical trials. The company has been at the forefront of adopting electronic data capture (EDC), mobile health (mHealth) technologies, and real-world evidence (RWE) solutions to optimize data collection and analysis.

Acquisition by Pamplona Capital Management:

In 2017, PAREXEL was acquired by Pamplona Capital Management, a private equity firm. The acquisition was completed through a merger agreement and made PAREXEL a privately held company.

PAREXEL continues to be a major player in the CRO industry, supporting pharmaceutical and biotechnology companies in the development of new drugs and medical devices. Its long history of expertise, global reach, and diverse service offerings have made it a significant partner in advancing medical research and drug development.

CHAPTER 11

My years working for KFORCE (2007-2011)

IN THE YEAR 2007, started working for another Clinical Research Organization known as Kforce Clinical Research. Kforce named by itself sounded like a group of superheroes. Its main importance was a CRO that worked for Pfizer and its cancer treatment drug known as Sutent. This drug named Sutent is now used to treat many types of cancers. At the time, I started working for Kforce as a Clinical Research Site Manager. I also took on the role as the Clinical Trials Site Selection Specialist in the Northeast region of the United States of America.

During this time for Pfizer clinical trials, Sutent was still in its infancy in the amount of cancer treatments it was available for. The clinical trials that I worked on during this time was vital for the eventual approval of Sutent to be used for various cancer treatments. Let us look at the background of Pfizer and its cancer treatment drug known as Sutent.

Pfizer is one of the world's largest and most well-known pharmaceutical companies, with a long and storied history. Here's an overview of the key milestones and developments in the history of Pfizer:

Founding and Early Years:

Pfizer was founded by cousins Charles Pfizer and Charles Erhart in Brooklyn, New York, in 1849. Originally, the company was a fine-chemicals business, producing substances like citric acid and santonin, a remedy for intestinal worms.

Mass Production of Medicines:

In the late 1800s and early 1900s, Pfizer started to focus on producing innovative pharmaceuticals. In 1950, Pfizer achieved a significant milestone by producing the world's first mass-produced penicillin, an antibiotic that revolutionized medicine and played a crucial role in treating bacterial infections during World War II and beyond.

Expansion and Diversification:

Throughout the 20th century, Pfizer expanded its operations and diversified its product portfolio through research and acquisitions. The company delved into various therapeutic areas, including cardiovascular, infectious diseases, central nervous system disorders, and more.

Development of Blockbuster Drugs:

Pfizer's commitment to research and development led to the creation of several blockbuster drugs that became household names. Notable examples include:

1. Lipitor (atorvastatin): A cholesterol-lowering statin medication that became one of the best-selling drugs in history.
2. Viagra (sildenafil): Originally developed as a treatment for hypertension and angina, Viagra gained fame for its use in treating erectile dysfunction.

Acquisitions and Mergers:

Pfizer has grown significantly through mergers and acquisitions. Notably, in 2000, Pfizer acquired Warner-Lambert, a major pharmaceutical company. This acquisition brought Lipitor into Pfizer's portfolio and solidified its position as a leading pharmaceutical company.

Other significant acquisitions include Pharmacia (2003) and Wyeth (2009), further expanding Pfizer's product offerings and global presence.

Research and Innovation:

Pfizer continues to invest heavily in research and development to discover new drugs and treatments. They collaborate with academic institutions, research centers, and biotech companies to explore cutting-edge therapies and technologies.

Vaccine Development:

Pfizer's history of vaccine development includes important contributions to public health. In partnership with BioNTech, Pfizer developed one of the first COVID-19 vaccines to receive emergency use authorization in late 2020.

Corporate Social Responsibility:

Pfizer is involved in various corporate social responsibility initiatives, including efforts to increase access to healthcare, philanthropic contributions, and programs to address global health challenges.

As a dynamic and evolving company, Pfizer's history is continuously unfolding with new discoveries, partnerships, and innovations.

Pfizer's Sutent (generic name: sunitinib) is a medication used for the treatment of certain types of cancer. Here's a brief overview of the history of Sutent:

Development:

Sutent was developed by Pfizer, a global pharmaceutical company, as a targeted therapy for cancer. It belongs to a class of drugs known as multi-targeted tyrosine kinase inhibitors (TKIs). These drugs work by interfering with specific proteins (tyrosine kinases) that play a role in the growth and spread of cancer cells.

FDA Approval:

Sutent received its first approval from the U.S. Food and Drug Administration (FDA) in January 2006. Initially, it was approved for the treatment of advanced renal cell carcinoma (kidney cancer) and gastrointestinal stromal tumor (GIST), which is a rare type of cancer that affects the digestive tract.

Expansion of Indications:

Over time, the FDA and other regulatory agencies around the world expanded the approved indications for Sutent to include other types of cancers. For instance, it has been approved for the treatment of advanced pancreatic neuroendocrine tumors (pNET) and certain types of advanced soft tissue sarcoma.

Clinical Trials and Research:

The development of Sutent involved extensive clinical trials to establish its safety and efficacy in treating various types of cancer. Clinical trials are essential to gather data on drug performance, side effects, and to identify potential new uses.

Impact and Further Research:

Sutent has been an important treatment option for patients with the approved cancer types, providing an alternative to traditional chemotherapy and improving outcomes for many individuals. Ongoing research and studies have likely continued after my last update to further explore its potential benefits and expand its applications.

CHAPTER 12

My years at MORPHOTEK (2012-2015)

AFTER COMPLETING MY monitoring oversight for the Sutent studies for Pfizer International, I then became a CRA Clinical Research Manager at Morphotek, Inc. in Exton, Pennsylvania.

Morphotek, Inc. is a biopharmaceutical company that specializes in the development of therapeutic monoclonal antibodies for the treatment of cancer and other diseases.

Here's an overview of the history of Morphotek:

Founding and Early Years:

Morphotek was founded in 2000 by Philip Sass, Dr. Nicholas C. Nicolaides, and Dr. Andrew Hiatt. The company was established as a spin-off from the research and development division of the University of Medicine and Dentistry of New Jersey (now part of Rutgers University).

Monoclonal Antibody Technology:

Morphotek focused on leveraging monoclonal antibody technology to discover and develop targeted therapies. Monoclonal antibodies are engineered proteins that can specifically recognize and bind to certain molecules, such as proteins or receptors found on the surface of cancer cells.

Growth and Research Focus:

In its early years, Morphotek primarily focused on research and development activities to discover and optimize monoclonal antibodies. The company aimed to create a diverse pipeline of therapeutic candidates for the treatment of various types of cancer and other diseases.

Partnerships and Collaborations:

Morphotek established collaborations and partnerships with other pharmaceutical and biotechnology companies to accelerate its research efforts and advance its drug development programs. These collaborations helped expand the company's capabilities and provided access to additional resources and expertise.

Eribulin Approval and Eisai Acquisition:

One of Morphotek's most significant achievements was the discovery and early development of Eribulin Mesylate (brand name: Halaven). Eribulin is a microtubule inhibitor used to treat metastatic breast cancer and certain types of soft tissue sarcoma. It was first approved by the U.S. Food and Drug Administration (FDA) for metastatic breast cancer in 2010.

In 2007, Eisai Co., Ltd., a Japanese pharmaceutical company, acquired Morphotek to strengthen its oncology portfolio. The acquisition allowed Eisai to gain access to Morphotek's innovative monoclonal antibody technology and drug development capabilities.

Continued Research and Development:

As a subsidiary of Eisai, Morphotek continued its research and development efforts, focusing on advancing Eribulin and other therapeutic candidates. The company remained dedicated to discovering and developing innovative treatments for cancer and other diseases.

Morphotek continued to be a subsidiary of Eisai, contributing to Eisai's global oncology pipeline and research initiatives. I recommend consulting up-to-date sources for the latest information on Morphotek and its contributions to the biopharmaceutical industry.

The clinical trials that I helped to manage showed some promise in monoclonal antibody technology. I would eventually continue with this type of technology with another company called Kite Pharmaceuticals. However after my time at Morphotek, I would first work on a different and very effective cancer treatment drug known as Keytruda.

CHAPTER 13

Merck and Keytruda Years (2015-2017)

IN 2015, I started to work in a monitoring and site management position for another CRO. PRA Health Sciences or Pharmaceutical Research Association is a leading contract research organization (CRO) that provides a wide range of services to support the biopharmaceutical and medical device industries in conducting clinical trials and drug development. The company offers comprehensive solutions throughout the entire drug development lifecycle, from early-stage development to post-approval studies. Let us have an overview of the company.

Founding and Early Years:

PRA Health Sciences was founded in 1982 by Dr. Kent Thoelke and Dr. Richard Staub as a small clinical research consulting firm in Virginia, USA. The company initially focused on providing expertise and support for clinical trial design and execution.

Growth and Expansion:

PRA Health Sciences experienced steady growth and expanded its services to meet the increasing demands of the pharmaceutical and biotechnology industries. The company established a global presence, opening offices in various countries to support clinical research on an international scale.

Diverse Service Offerings:

PRA Health Sciences offers a comprehensive range of services to support clinical trials, including clinical trial management, site selection and monitoring, data management, biostatistics, medical writing, regulatory consulting, safety and pharmacovigilance, and real-world evidence (RWE) research.

Strategic Acquisitions:

PRA Health Sciences pursued strategic acquisitions to expand its capabilities and geographic reach. Notable acquisitions include Symphony Health Solutions, which strengthened PRA's real-world data and analytics capabilities, and Symphony Clinical Research, a specialized provider of in-home and alternate-site clinical services.

IPO and Public Listing:

In 2014, PRA Health Sciences completed its initial public offering (IPO) and became a publicly-traded company, listing its shares on the NASDAQ stock exchange under the ticker symbol "PRAH." The IPO provided the company with additional capital for further expansion and investment in research and technology.

Innovative Technologies and Solutions:

PRA Health Sciences has been at the forefront of adopting innovative technologies and solutions to enhance clinical trial processes. The company has invested in electronic data capture (EDC), mobile health (mHealth) technologies, wearables, and other digital solutions to optimize data collection and analysis.

Global Impact:

With its extensive global network of research sites and collaborations with pharmaceutical and biotechnology companies worldwide, PRA Health Sciences has played a significant role in advancing medical research and facilitating the development of new treatments and therapies for various diseases.

PRA Health Sciences continues to be a major player in the CRO industry, providing essential support and expertise to help bring new medical advancements to patients. I recommend consulting up-to-date sources for the latest information on PRA Health Sciences and its contributions to clinical research and drug development.

After a few months in training for PRA in 2015, I was assigned to the Keytruda study drug program. Keytruda is a cancer treatment drug that has proven to be very effective. Merck was testing this drug on various types of cancers.

Merck & Co., Inc., commonly known as Merck, is one of the oldest and largest pharmaceutical companies in the world. The company has a rich history dating back to the 19th century. Here's an overview of the key milestones and developments in the history of Merck:

Merck was founded in 1668 by Friedrich Jacob Merck in Darmstadt, Germany. It started as a small pharmacy, and the Merck family continued to run the business for generations.

Expansion to the United States:

In 1887, George Merck, a descendant of the original founder, immigrated to the United States and established Merck & Co. in New York City. The American subsidiary began as a small sales office and gradually grew into a significant presence in the U.S. pharmaceutical industry.

Mass Production of Drugs:

During the early 20th century, Merck played a pivotal role in the mass production of important drugs. In 1913, Merck scientists developed the first U.S. Food and Drug Administration (FDA)-approved vaccine for diphtheria, a deadly infectious disease at the time. They also played a role in the production of penicillin during World War II, which significantly increased the availability of this life-saving antibiotic.

Key Medical Breakthroughs:

Merck made several significant medical breakthroughs throughout its history. In the 1950s, the company developed streptomycin, one of the first effective treatments for tuberculosis. In 1952, Merck introduced methyldopa, a medication for hypertension. The company also contributed to the development of the MMR vaccine (measles, mumps, rubella) and the HPV vaccine, Gardasil.

Expansion and Mergers:

Over the years, Merck expanded its operations globally and pursued mergers and acquisitions to strengthen its position in the pharmaceutical industry. In 1953, Merck acquired the Philadelphia-based Sharp & Dohme, adding a well-established name in the pharmaceutical market to its portfolio.

In 2009, Merck completed a significant merger with Schering-Plough, a multinational pharmaceutical company. The merger broadened Merck's product range and therapeutic areas, making it one of the largest pharmaceutical companies worldwide.

Continued Innovation and Research:

Merck has maintained a strong commitment to research and development. The company invests heavily in discovering and developing new drugs and treatments for various medical conditions. Over the years, Merck's research efforts have led to the development of medicines for cardiovascular diseases, diabetes, infectious diseases, cancer, and more.

Corporate Social Responsibility:

Merck has been actively involved in various corporate social responsibility initiatives, including efforts to improve access to healthcare and support global health programs.

Merck continues to be a major player in the pharmaceutical industry, with a focus on innovative therapies, research, and development. The company's history is a testament to its commitment to advancing medicine and improving global health.

In 2015, Keytruda was a main study drug in the Merck program. Keytruda is a brand name for the medication Pembrolizumab, which is an immune checkpoint inhibitor used in cancer immunotherapy. Developed by Merck & Co., Inc., Keytruda is one of the pioneering drugs in the field of immunotherapy and has shown significant success in the treatment of various types of cancer.

Mechanism of Action:

Keytruda works by blocking a specific immune checkpoint known as PD-1 (programmed cell death protein 1). PD-1 is a protein found on the surface of certain immune cells, such as T cells. Its primary role is to prevent the immune system from attacking healthy cells in the body and to maintain immune self-tolerance.

In cancer, tumors can exploit the PD-1 pathway to evade the immune system. They express a ligand called PD-L1 (programmed death-ligand 1) on their surface, which binds to PD-1 on immune cells, effectively "switching off" the immune response against the cancer cells. By blocking the interaction between PD-1 and PD-L1, Keytruda helps reactivate the immune system, allowing it to recognize and attack cancer cells.

Indications:

Keytruda has received approval for the treatment of several types of cancer, both as a monotherapy and in combination with other therapies. As of my last update in September 2021, some of the approved indications for Keytruda included:

1. Advanced Melanoma: Keytruda was one of the first immunotherapies approved for advanced melanoma, a type of skin cancer.
2. Non-Small Cell Lung Cancer (NSCLC): Keytruda is used for the treatment of certain types of NSCLC, particularly those with high levels of PD-L1 expression.
3. Head and Neck Squamous Cell Carcinoma (HNSCC): Keytruda is approved for recurrent or metastatic HNSCC that has progressed after prior chemotherapy.
4. Classical Hodgkin Lymphoma: Keytruda is indicated for classical Hodgkin lymphoma that has relapsed or failed to respond to previous treatments.
5. Urothelial Carcinoma: Keytruda is used to treat advanced urothelial carcinoma, which

includes bladder cancer, for patients who have previously received platinum-containing chemotherapy.
6. MSI-High or dMMR Tumors: Keytruda is approved for the treatment of tumors with microsatellite instability-high (MSI-H) or mismatch repair deficiency (dMMR), regardless of the tumor's location.

Ongoing Research and Expanding Indications:

Immunotherapy, including Keytruda, is an area of active research and ongoing clinical trials. The drug's indications may continue to expand to include other types of cancer, and researchers are investigating its potential in combination therapies and in earlier stages of cancer treatment.

There may have been further developments regarding Keytruda since then. Always consult a healthcare professional or updated sources for the most current information on specific medications and treatments for Keytruda.

CHAPTER 14

Chiltern, Covance, and Labcorp Years (2017-2022)

IN 2017, I took the opportunity to work as a Clinical Research Associate (Level III) at another CRO named Chiltern at the time. I worked for Kite Pharmaceutical studies. The study drugs were targeted MRNA medications designed to fight various forms of cancer. Chiltern eventually was bought out by Covance. The company named Covance is a prominent Contract Research Organization (CRO) with a rich history in providing a wide range of services to support drug development and clinical research for pharmaceutical and biotechnology companies.

Here's an overview of the history of Covance:

Founding and Early Years:

Covance was founded in 1968 in Madison, Wisconsin, USA, by Dr. Wallace Calvin and Joe Herring. Originally known as Hazleton Laboratories, the company initially focused on providing toxicology and safety testing services for pharmaceutical, chemical, and agrochemical industries.

Growth and Expansion:

During the 1970s and 1980s, Covance expanded its services and client base, becoming a significant player in the preclinical research and toxicology testing space. The company established additional facilities in the United States and globally to meet the growing demand for its services.

Name Change to Covance:

In 1997, Hazleton Laboratories changed its name to Covance, reflecting its evolving focus on providing comprehensive drug development services, including early-stage development, clinical trials, and central laboratory services.

Entry into Clinical Research and Drug Development:

With the name change to Covance, the company expanded its capabilities into clinical research and drug development services. Covance developed expertise in managing clinical trials, data management, site monitoring, regulatory support, and other essential aspects of clinical research.

Growth through Acquisitions:

Covance pursued strategic acquisitions to expand its capabilities and geographical presence. Notable acquisitions include acquiring Quintiles' central laboratory services business in 2011, which further solidified Covance's position as a leading central laboratory services provider.

Merger with LabCorp:

In 2014, Covance and Laboratory Corporation of America Holdings (LabCorp), a leading global life sciences company, announced a definitive agreement for LabCorp to acquire Covance. The acquisition was completed in early 2015. The merger brought together Covance's drug development and clinical research expertise with LabCorp's diagnostic and laboratory services capabilities, creating a comprehensive life sciences company.

Continued Innovations and Services:

As part of LabCorp, Covance continued to innovate and expand its service offerings. The company has been at the forefront of adopting technologies such as biomarkers, real-world evidence (RWE), and precision medicne approaches to support drug development and personalized healthcare.

Global Presence and Impact:

Covance has a significant global presence, with operations in numerous countries and collaborations with pharmaceutical and biotechnology companies worldwide. The company's services have played a vital role in advancing medical research and facilitating the development of new drugs and medical treatments.

As previously stated, I was a CRA Site Manager for primarily the Kite cancer treatment program. This program proved very intense as the study drug was proving to be very effective. The company was a smaller company with the heavy burden of moving forward a study drug that used advanced technology that could have a huge impact on cancer treatment. Kite Pharmaceuticals is a biopharmaceutical company that specializes in the development of innovative cancer immunotherapies. Let us have an overview of the history of Kite Pharmaceuticals.

Founding and Early Years of Kite Pharmaceuticals:

Kite Pharmaceuticals was founded in 2009 by Arie Belldegrun, M.D., and Joshua Kazam, Ph.D., with the goal of harnessing the power of the immune system to develop transformative cancer therapies. The company's name, "Kite," was inspired by a sailboat kite, symbolizing the sense of freedom and movement they hoped to bring to cancer patients through their therapies.

CAR T-Cell Therapy Development:

Kite Pharmaceuticals became a pioneer in the field of chimeric antigen receptor (CAR) T-cell therapy. This innovative approach involves genetically modifying a patient's own T cells to express CARs, which are engineered receptors that recognize and target specific cancer cells.

Early clinical trials in the mid-2010s showed promising results for CAR T-cell therapy, particularly in treating certain types of blood cancers, such as aggressive non-Hodgkin lymphomas and acute lymphoblastic leukemia (ALL).

FDA Approvals:

In October 2017, Kite Pharmaceuticals achieved a significant milestone when the U.S. Food and Drug Administration (FDA) granted approval to its CAR T-cell therapy named Yescarta (axicabtagene ciloleucel). Yescarta became the second CAR T-cell therapy to receive FDA approval, following Novartis' Kymriah.

Yescarta was approved for the treatment of adults with certain types of large B-cell lymphoma, including diffuse large B-cell lymphoma (DLBCL) that had relapsed or was refractory to two or more previous lines of therapy. The approval marked a major advancement in personalized cancer treatment.

Acquisition by Gilead Sciences:

In August 2017, shortly after the FDA approval of Yescarta, Kite Pharmaceuticals was acquired by Gilead Sciences, a leading biopharmaceutical company. The acquisition further strengthened Gilead's presence in the oncology field and expanded its portfolio to include cutting-edge CAR T-cell therapies.

Ongoing Research and Expansion:

Following the acquisition by Gilead Sciences, Kite Pharmaceuticals continued its research and development efforts in CAR T-cell therapy and other cancer immunotherapies. The company aimed to expand the indications for Yescarta and develop new CAR T-cell therapies to address other types of cancer.

There may have been further advancements or updates regarding Kite Pharmaceuticals since the writing of this book. Always consult updated sources for the latest information on specific companies and their contributions to medical science and cancer treatment.

CHAPTER 15

Approval of Yescarta

THE BIGGEST ACCOMPLISHMENT I had during my years working in the Kite program was the approval of the Yescarta study drug. Yescarta (axicabtagene ciloleucel) is a groundbreaking cancer therapy developed by Kite Pharmaceuticals, a subsidiary of Gilead Sciences. Yescarta is a type of CAR T-cell therapy, which stands for chimeric antigen receptor T-cell therapy.

Mechanism of Action:

Yescarta is designed to treat certain types of aggressive non-Hodgkin lymphomas, particularly diffuse large B-cell lymphoma (DLBCL) and primary mediastinal large B-cell lymphoma (PMBCL). It is used in adult patients who have not responded to, or have relapsed after, two or more prior lines of treatment.

The therapy involves a complex process that starts by collecting a patient's own T cells, a type of immune cell responsible for fighting infections and cancer. These T cells are then genetically modified in the laboratory to express chimeric antigen receptors (CARs) on their surface.

CARs are engineered receptors that can recognize and bind to a specific protein called CD19, which is found on the surface of B cells, including cancerous B cells in lymphomas. Once the T cells are modified to express the CARs, they are expanded in large numbers and then infused back into the patient's bloodstream.

When the CAR T cells encounter CD19-expressing cancer cells, they become activated and initiate a potent and targeted immune response. They multiply rapidly and attack the cancer cells, leading to their destruction.

Approval and Impact:

Yescarta received accelerated approval from the U.S. Food and Drug Administration (FDA) in October 2017, making it one of the first commercially available CAR T-cell therapies. It marked a significant milestone in the field of cancer immunotherapy, as Yescarta was the second CAR T-cell therapy to receive FDA approval, after Novartis' Kymriah.

Yescarta's approval was based on the results of the ZUMA-1 clinical trial, which demonstrated impressive response rates and durable remissions in patients with refractory or relapsed DLBCL. The therapy has shown the potential to provide a new treatment option for patients with limited alternatives and has offered hope to those facing aggressive and difficult-to-treat forms of lymphoma.

Continued Research and Development:

Following the approval of Yescarta, research and development in the field of CAR T-cell therapy have continued to advance. Studies are ongoing to explore the therapy's potential in other types of cancer and to optimize its efficacy and safety. Additionally, efforts are being made to address challenges related to manufacturing and the high costs associated with CAR T-cell therapies.

As with any medical treatment, Yescarta is not without risks and side effects, including cytokine release syndrome (CRS) and neurologic toxicities. However, ongoing research and clinical experience have provided valuable insights into managing and minimizing these adverse events. Another major hurdle is the current cost of CAR T-cell treatment which can be very expensive to the patient over time of use.

I recommend consulting the latest sources and medical professionals for the most current information on Yescarta and other CAR T-cell therapies.

CHAPTER 16

Reasons and Importance of Clinical Research Monitoring

I WANT TO conclude this book by discussing the importance of monitoring and clinical research trials. Clinical research monitoring is a critical component of ensuring the integrity, quality, and ethical conduct of clinical trials. Here are some key reasons why clinical research monitoring is important:

1. **Participant Safety:** Monitoring ensures that participants are protected from potential risks associated with the trial. Monitoring helps

identify any adverse events or side effects early on, allowing for prompt intervention and ensuring participant safety.

2. **Data Quality**: Monitoring ensures the accuracy, reliability, and integrity of the data collected during the trial. By verifying that data is collected according to the study protocol and regulatory requirements, monitoring helps maintain the quality of the data, which is essential for drawing valid conclusions.
3. **Protocol Compliance**: Monitoring ensures that the trial is conducted in accordance with the approved study protocol, which outlines the objectives, methodology, and procedures of the trial. Monitoring helps identify any deviations from the protocol and ensures that the trial is conducted consistently and ethically.
4. **Regulatory Compliance**: Monitoring ensures that the trial adheres to relevant regulatory requirements, such as those outlined by government agencies like the Food and Drug Administration (FDA) in the United States or the European Medicines Agency (EMA) in Europe. Compliance with regulations is essential for

obtaining regulatory approval for new drugs or medical devices.

5. **Data Integrity:** Monitoring helps prevent data fraud or misconduct by verifying the accuracy and authenticity of the data collected during the trial. This is crucial for maintaining the credibility and trustworthiness of the trial results.

6. **Early Detection of Issues:** Monitoring allows for the early detection of any issues or challenges that may arise during the trial, such as problems with recruitment, data collection, or site performance. Early intervention can help address these issues promptly and mitigate their impact on the trial's outcomes.

7. **Efficient Resource Utilization:** Monitoring helps optimize the use of resources, including time, personnel, and funding, by identifying areas where improvements can be made or where resources may be allocated more effectively.

8. **Trial Success:** Ultimately, effective monitoring contributes to the overall success of the clinical trial by ensuring that it is conducted

in a rigorous, compliant, and ethical manner, leading to reliable results that can advance medical knowledge and improve patient care.

In summary, clinical research monitoring plays a vital role in safeguarding the safety of the participant in a clinical trial. Clinical research monitoring also maintains data quality, data integrity, protocol compliance, and regulatory compliance. Clinical research monitoring also detects issues early while ultimately contributing to the success of clinical trials. I hope this book has provided you with a good background knowledge on this important clinical research role. Hopefully, this book will provide you with the background and incentive to appreciate this job position. I hope some of you are even inspired to eventually join us in this field of clinical research. Continued medical research leading to health of the human population can help contribute to the growth of humanity in our near future.

REFERENCES

MY CAREER IN CLINICAL RESEARCH

Included are Biographic along with Photographic Books

1. Abell, D.F. and Hammond, J.S. *Strategic Market Planning: Problems and Analytical Approaches.* Englewood Cliffs, N.J.: Prentice Hall. 1979.
2. Amdur, Robert J. and Bankert, Elizabeth A. *Institutional Review Board: Management and Function.* Sudbury, M.A.: Jones and Bartlett Publishers. Second Edition. 2006.

3. Aranha, Hazel and Vega-Mercado, Humberto. *Handbook of Cell and Gene Therapy.* Boca Raton, F.L.: CRC Press. 2023.
4. Bolden, Don. *LabCorp: The DNA of a Corporation.* Independent. 2019.
5. Emanuel, Ezekiel J. and Moreno, Jonathan D. *Ethical and Regulatory Aspects of Clinical Research.* Baltimore, M.D.: Johns Hopkins University Press. 2003.
6. Graeber, Charles. *The Breakthrough: Immunotherapy and the Rush to Cure Cancer.* Twelve Publishing. 2018.
7. General Books LLC. *Pharmaceutical Companies of France: Sanofi Aventis.* General Books LLC. 2010.
8. Hutchins, Amey A. *University of Pennsylvania (PA): Campus History Series.* Mount Pleasant, S.C.: Arcadia Publishing. 2004.
9. Rodengen, Jeffrey L. *The Legend of Pfizer.* Write Stuff Enterprises Inc. 1999.
10. United States Food and Drug Administration. *2006 Code of Federal Regulations and ICH Guidelines. Good Clinical Practice Reference Guide Plus European Union Clinical Trials*

Directive. United States Food and Drug Administration. 2006.
11. Vagelos, P. Roy. *Medicine, Science, and Merck.* Cambridge, M.A.: Cambridge University Press. 2004.

General search engines on the internet can be used such as Google and Microsoft Start for search of official websites to pharmaceutical companies and organizations. History of PAREXEL, Morphotek Incorporated and Eisai Pharmaceuticals can be found through general internet search.

www.ingramcontent.com/pod-product-compliance
Lightning Source LLC
Chambersburg PA
CBHW020451220526
45464CB00002B/953